DATE DUE

Amazing Planet Earth

MASSIVE MOUNTAINS

TERRY JENNINGS

A+

Smart Apple Media

$28.50

Smart Apple Media
P.O. Box 3263
Mankato, MN 56002

Printed in the United States of America

Library of Congress Cataloging-in-Publication Data

Jennings, Terry J.
   Massive mountains / Terry Jennings.
      p. cm. -- (Amazing planet earth)
   Includes index.
   ISBN 978-1-59920-370-6 (hardcover)
   1. Mountains--Juvenile literature. I. Title.
   GB512.J44 2010
   551.43'2--dc22

                              2008055497

Created by Q2AMedia
Book Editor: Michael Downey
Art Director: Rahul Dhiman
Designers: Ritu Chopra, Ranjan Singh
Picture Researcher: Shreya Sharma
Line Artist: Sibi N. Devasia
Coloring Artist: Mahender Kumar

All words in **bold** can be found in the glossary on pages 30–31.

Web site information is correct at time of going to press. However, the publishers cannot
accept liability for any information or links found on third-party web sites.

Picture credits
t=top b=bottom c=center l=left r=right
Cover Image: Alessio Ponti/ Shutterstock.
Back Cover: shutterstock.

Insides: Nathan Jaskowiak/ Shutterstock: Title Page, Nathan Jaskowiak/ Shutterstock: 4-5, Jorg Jahn/ Shutterstock: 6, Associated Press:
7t, Steffen Foerster Photography/ Shutterstock: 8, National Geographic/ Getty Images: 9, iStockphoto: 12, EarthObservatory/ NASA:
13, Ilya D. Gridnev/ Shutterstock: 15b, Cristina/ Shutterstock: 16, Wolfgang Amri/ Shutterstock: 17, Michael Shake/ Shutterstock: 18,
Karen Kasmauski/ Corbis: 19, Amygdala Imagery/ Shutterstock: 20, Urosr/ Shutterstock: 21, Dwi Oblo / Reuters: 22, Associated Press:
23, David P. Lewis/ Shutterstock: 25, Corbis Sygma: 26, Bmflv Minich/ Associated Press: 27, Steve Estvanik/ 123rf: 28, Jorg Jahn/
Shutterstock: 31.

Q2AMedia Art Bank: 7b, 10, 11, 14, 15t, 24,

9 8 7 6 5 4 3 2 1

Contents

Beauty and Danger

Mountains are some of the most beautiful places on Earth. But they can also be extremely dangerous. Freezing temperatures and high winds can be deadly for even the most experienced climbers.

Magic Mountains

Mountains make up a quarter of Earth's surface. There are even mountain **ranges** under the world's oceans. All of the world's largest rivers begin life in the mountains, and mountains have a major effect on the weather. Forestry companies grow trees on mountain slopes for wood. Mining companies dig out valuable minerals from mountains.

• The Andes are the world's longest mountain chain. They stretch 5,500 miles (8,900 km) along the western side of South America.

Life and Death

As soon as a mountain is formed, it begins to wear away. Little by little, the mountain crumbles. The pieces are carried away by the wind, rivers, and **glaciers**. Eventually, they reach the oceans. These changes in mountains usually take place very slowly. But sometimes, mountains change with terrifying speed and put the lives of people who live and work near them in danger.

DATA FILE

- Mountains are in 75 percent of the world's countries.

- Mountains are made of solid **rock**. Hills can be made of solid rock or rock fragments built up by glaciers or wind.

- More than half of all fresh water comes from mountains.

- One in ten of the world's population live near mountains.

- Mountain **peaks** are always very cold and often covered in snow.

The Himalayas

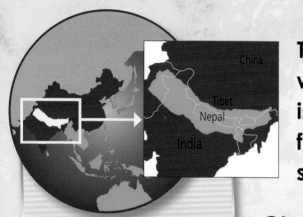

China

Tibet
Nepal

India

Name: The Himalayas
Location: Asia
Length: 1,550 mi
(2,500 km)
Age: 30–50 million years
Mountain type: Fold
Highest peak: Mount
Everest, 29,030 ft (8,848 m)

The Himalayas contain many of the world's highest peaks. The tallest is Mount Everest. These mountains form a gigantic barrier of rock that separates India and Nepal from Tibet.

Giant Humps

Millions of years ago, the Himalayas were not mountains. They were flat areas of land at the bottom of a shallow ocean. The Himalayas began to grow about 50 million years ago when two sections of Earth's **crust** started crunching against each other. This forced the layers of rock up to form giant humps and folds that we now know as the Himalayas.

• Mount Everest lies among the snow-covered peaks of the Himalayas.

• In 1953, New Zealander Edmund Hillary and Tenzing Norgay from Nepal were the first people to reach the top of Everest.

Monsoon Rains

For four months each year, monsoon winds blow north over the Himalayas. These cause heavy rain and snow to fall on the southern side of the Himalayas. This is why there are lush tropical forests and snow-covered peaks on the southern side. By the time the winds reach the northern side of the Himalayas, they have lost most of their moisture. On these north-facing slopes, there is dry desert.

• Monsoon winds release most of their rain on the southern slopes of the Himalayas.

Heavy rain

Dry winds

Parched desert

Wet winds

Not Enough Wood

Several big rivers are fed by the ice and snow that cover much of the Himalayas. Hundreds of millions of people in India, Pakistan, and Bangladesh get their water from these rivers. But the Himalayas are changing. In the last 30 to 40 years, the population of Nepal, on the southern slopes of the Himalayas, has doubled. There is now not enough food and firewood to go around. More than two-thirds of the trees have been cut down for firewood and to make new fields.

Overflowing Rivers

Normally, tree roots soak up water and hold soil in place. Now, with fewer trees in the Himalayas, soil is washed into the rivers when it rains. This makes the rivers overflow, causing flooding in the **valleys** and plains across India and Bangladesh. In Nepal, entire villages have been swept away by torrents from overflowing rivers.

- In Nepal, the lower slopes of the Himalayas have been cut into steps to stop soil from being washed away by overflowing rivers.

Litter Problem

A big problem on Mount Everest and other Himalayan mountains is the litter that is left by the many climbing expeditions and tourists that visit the area. Over the years, a huge quantity of litter has collected. Because it is so cold, the litter does not rot or rust away.

DATA FILE

- According to one report, there are at least 220 tons (200 t) of litter on the higher slopes of Everest, making it the world's highest garbage dump.

- Mountaineers have left behind more than 50,000 glass bottles, ropes, tents, food packaging, and even a wrecked helicopter.

- In 1998, the Nepalese government imposed a ban on taking bottled drinks to the Mount Everest region.

- Many gas and oxygen canisters have been dumped by mountain climbers on Mount Everest.

Fold and Block Mountains

Earth's crust is made up of huge slabs of rock called plates. These move very slowly, sometimes crashing into each other, sometimes moving apart. The movement of these plates has caused all mountains to form.

Fold Mountains

Earth's **plates** float on the very hot rocks below them and make very small movements each year. When two of these plates crash, the layers of rock may be pushed up and folded over each other, like a giant rumpled blanket. This folding can make several long rows of mountains that are called **fold mountains**. The highest mountains are in the middle, with smaller mountains on either side, and rows of hills beyond. The Himalayas, the European Alps, and the U.S. Appalachians are all fold mountains.

Fold mountain

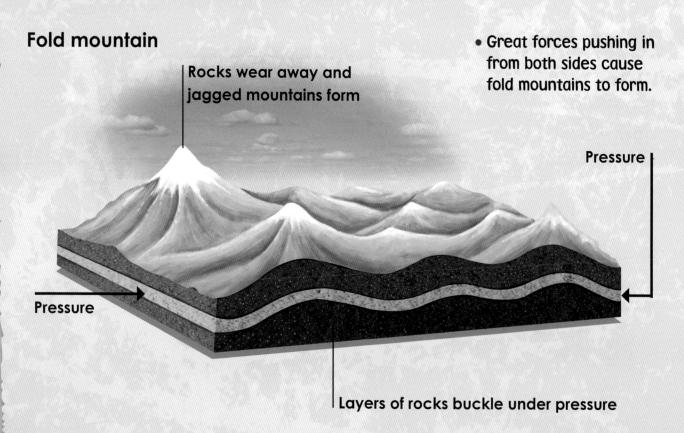

Rocks wear away and jagged mountains form

• Great forces pushing in from both sides cause fold mountains to form.

Pressure

Pressure

Layers of rocks buckle under pressure

Block Mountains

As Earth's plates push and pull against each other, rocks sometimes crack under the strain. These cracks are called faults. When two faults are close together, the chunk of Earth's crust between them can sometimes collapse to form a rift valley. Or, if the plates shove together, they may squeeze a great slab of rock up. The raised parts are called **block mountains**. The Sierra Nevada mountains (in California and Nevada) and the Harz mountains (in Germany) are block mountains. They were formed long ago during some of the world's largest earthquakes.

Block mountain

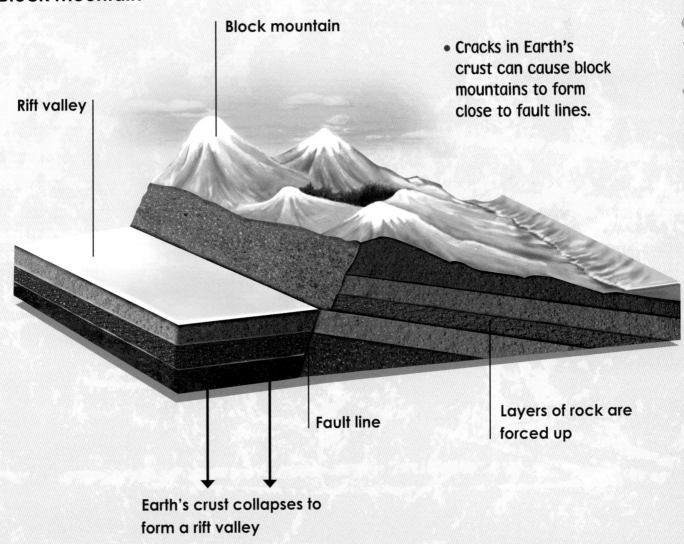

Block mountain

Rift valley

• Cracks in Earth's crust can cause block mountains to form close to fault lines.

Fault line

Layers of rock are forced up

Earth's crust collapses to form a rift valley

Mount Kilimanjaro

Mount Kilimanjaro is 19,340 feet (5,895 m) high. This extinct volcano is the world's highest freestanding mountain. It is also the tallest mountain in Africa.

Steam and Sulfur Dioxide

Mount Kilimanjaro is the largest **volcano** in a belt of about 20 volcanoes in East Africa. Kilimanjaro formed between about 1.8 million and 10,000 years ago. Although Mount Kilimanjaro is an extinct volcano, steam and **sulfur dioxide** still stream from it. Even though the mountain is very near to the equator, until recently its summit always had a covering of snow and ice.

- Snow covers the top of Tanzania's Mount Kilimanjaro for a short time each year.

Name: Mount Kilimanjaro
Location: Tanzania, Africa
Height: 19,340 ft (5,895 m)
Age: 10,000 to 1.8 million years
Mountain type: Volcano

Disappearing Ice

During the 1900s, Mount Kilimanjaro's ice cap shrunk dramatically. In 1912, the ice covered an area measuring more than 4 square miles (10 sq km). Now four-fifths of the ice cap has disappeared, leaving just small patches. Scientists think that the ice at the top of Kilimanjaro will be completely gone by 2020. Then there will not be enough **meltwater** to keep rivers and springs running all year round.

• The ice cap on the summit of Mount Kilimanjaro is shrinking fast.

News Flash

March 15, 2005

The snowcapped summit of Mount Kilimanjaro has melted away to reveal the tip of the African peak for the first time in 11,000 years. The glaciers and snow, which kept the summit white, have almost completely disappeared. Scientists had predicted the melt. However, it is 15 years sooner than they had estimated. The white peak of the 19,340 foot (5,895 m) mountain has long formed a stunning part of Tanzanian landscape.

Volcanic and Dome Mountains

Below Earth's crust is a layer of rock called the mantle. Parts of the mantle are so hot that the rock has melted to form a gooey substance called magma. When magma reaches the surface, it can form spectacular mountains.

Volcanic mountain

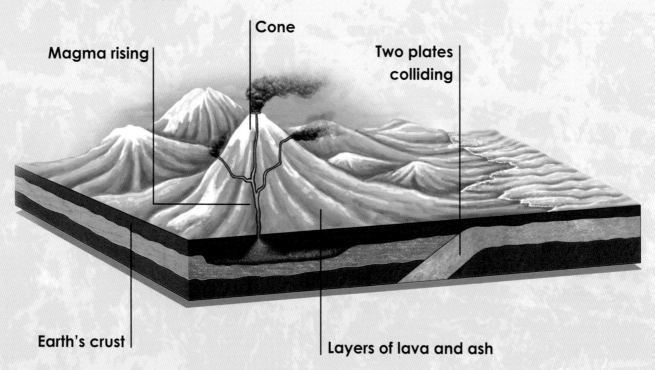

Cone

Magma rising

Two plates colliding

Earth's crust

Layers of lava and ash

Volcanic Mountains

A volcanic mountain forms when molten rock, or **magma**, from deep inside Earth escapes to the surface. When it reaches the surface, the magma is called lava. Most volcanoes are found at weak points in Earth's crust where two or more plates crash together or move apart. Like Mount Kilimanjaro, Mount St. Helens in Washington and Mount Pinatubo in the Philippines were also formed by volcanoes.

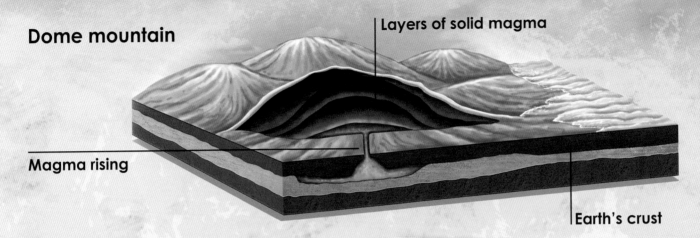

Dome mountain

Layers of solid magma

Magma rising

Earth's crust

Dome Mountains

When magma pushes its way upward, it may collect just beneath Earth's surface like a gigantic blister. The magma cools into solid rock and forms a **dome mountain**. Unlike a volcano, the magma under a dome mountain does not have enough force to become a volcano. Navajo Mountain in the southwest United States and Showashinzan in Japan are examples of dome mountains.

- Showashinzan in Japan is a dome mountain. It is like a gigantic blister filled with solid magma.

DATA FILE

- Earth's surface is made up of seven major plates and many minor plates.

- Plates move up to 5 inches (13 cm) a year.

- A plate may travel 31 miles (50 km) in a million years.

- The Andes in South America and the Rocky Mountains in the United States lie close to the edges of major plates.

European Alps

The Alps are the largest mountain chain in Europe. They stretch 620 miles (1,000 km) from Austria and Slovenia in the east to Germany and France in the west.

Huge Valleys and Lakes

The European Alps, which formed between 33 and 15 million years ago, have several peaks more than 13,000 feet (4,000 m) high. At 15,771 feet (4,807 m), Mont Blanc in France is the highest peak. Over the last 2 million years, the mountain landscape has been changed by glaciers. These rivers of ice have carved out deep U-shaped valleys and huge lakes, such as Lake Como and Lake Garda.

Name: European Alps
Location: Southcentral Europe
Length: 620 mi (1,000 km)
Age: 15–33 million years
Mountain type: Fold
Highest mountain: Mont Blanc 15,771 feet (4,807 m)

• About 20,000 mountaineers climb France's Mont Blanc every year.

Skiing, Hiking, and Biking

The Alps are very popular for sightseeing and sports. During the winter, people ski, snowboard, toboggan, and snowshoe. In summer, the Alps are popular with hikers, mountain bikers, paragliders, and mountaineers. Many of the lakes attract swimmers, sailors, and windsurfers. The lower areas and larger towns of the Alps are easy to reach by highways and roads. To make it easier to get around, tunnels have been cut through the Alps to allow travel at all times of the year.

• From December to April, skiing is a popular sport in the European Alps.

DATA FILE

• Many of Europe's large rivers, including the Rhine and the Danube, begin in the Alps.

• Alpine waterfalls and rivers are used to generate electricity.

• Below the snowline, pastures are used for grazing cattle and sheep in summer.

• In Alpine valleys and foothills, crops are planted. Grapes are grown on some sunny slopes.

• In 1991, the body of a man was found in ice in the Alps. A bow, arrows, and a copper axe were nearby. Scientists discovered the man had died 5,200 years ago!

Appalachian Mountains

The Appalachian mountain range runs along the east coast of North America from the province of Quebec in Canada to Alabama in the southern United States.

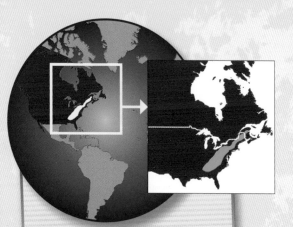

Name: Appalachian Mountains
Location: Atlantic coast of North America
Length: 1,490 mi (2,400 km)
Age: 300 million years
Mountain type: Fold
Highest mountain: Mount Mitchell, North Carolina, 6,683 ft (2,037 m)

Forest and Farmland

The Appalachians are low and gentle with an average height of only 3,000 feet (900 m). They are special, however, because they cover an area about 1,490 miles (2,400 km) long and at times more than 250 miles (400 km) wide. Much of the range is covered by dense forest. The Great Appalachian Valley, which runs the length of the mountains, is extremely fertile farmland. The Appalachians are popular with hikers. The Appalachian Trail is 2,200 miles (3,500 km) from Maine to Georgia and takes up to six months to complete!

- Much of the Appalachian Mountain range is covered in thick forest.

Rocks and Minerals

Not all parts of the Appalachians are beautiful, however. The mountains contain huge deposits of valuable rocks and minerals. These include iron ore, coal, slate, limestone, oil, and gas. To get coal, about 470 mountains have been blown up and leveled in recent years. Before the coal can be mined, the forest is bulldozed. The topsoil is scraped away and explosives are set off in the rocks below. The rubble is then tipped into the valleys. More than 7,000 valleys have been filled and some 685 miles (1,100 km) of rivers and streams have disappeared under rubble or have been polluted with toxic waste.

- In the Appalachians, miners have sliced off the tops of mountains to reach mineral deposits.

DATA FILE

- The Appalachians are one of the oldest mountain chains on Earth.

- The name comes from the Apalachee tribe, who were the first inhabitants of the area.

- The Appalachian Mountain range does not have any volcanoes.

- Many Appalachian streams are used to make hydroelectric power.

- Farms and orchards are found in the valley bottoms. Potatoes and wheat are grown in the north. In the south, farmers grow corn and tobacco and raise poultry.

Weathering and Erosion

Mountains may seem solid and permanent, but they begin to crumble almost as soon as they rise. Weaknesses appear in the rocks that let in air and water. This starts a process called weathering.

Breaking Rocks

When water gets into small cracks in rocks, it makes the cracks wider. When water freezes, it expands as it turns into ice. Eventually, pieces of rock break off and slide down a mountain's steep slopes. This loose rock is called **scree**. Wind, rivers, and glaciers carry a lot of the rock fragments away from the mountains. Most of it ends up at the bottom of the sea.

• At 14,495 feet (4,418 m), craggy Mount Whitney is the highest mountain in the Sierra Nevadas.

Jagged to Rounded

As pieces of rock break away from mountains, jagged peaks are left. Over time, however, even these jagged peaks are worn away to leave softly rounded mountain tops. The Himalayas, the Andes, and the European Alps still have jagged peaks that have not eroded very much. The Appalachians in the United States and Australia's Uluru, or Ayers Rock, have been worn down over millions of years.

DATA FILE

- No movement is involved in **weathering**, but when the loosened rock material starts to move due to wind, moving water, or ice, it is called **erosion**.

- The world's rivers carry 22 billion tons (20 billion t) of loose rock fragments, or sediment, to the oceans each year.

- All mountains slowly erode. One result of this is that they gradually lose weight. When this happens, Earth's crust slowly pushes the mountain upward.

• Uluru, or Ayers Rock, in Australia is only 1,142 feet (348 m) high. It was formed 300 million years ago and was once much higher but has slowly worn away.

Landslides in Java

Landslides are common in Indonesia. But in December 2007, days of torrential monsoon rains and flooding caused the most devastating landslides in the area for 25 years.

Name: Java landslides
Location: Java, Indonesia
Type of disaster: Landslide
Deadly event: December 2007
Fatalities: At least 120

Java at Risk

Landslides often occur in Indonesia during the rainy season. As rain batters the sides of a mountain, it soaks the ground. This causes huge chunks of earth to break away. The inhabitants of the island of Java are always at risk because so many people live on the lower slopes of mountains and in the valleys. The removal of the trees from muddy mountain slopes has worsened the problem.

- Rescuers search for survivors in the mud left by the Java landslides.

Deadly Mud

In December 2007, many days of torrential rain caused severe flooding and landslides across Java. The flooding was made worse by extra high ocean tides. Tens of thousands of people were left homeless. They had to wade through chest-high water, clutching their belongings above their heads. The worst single incident was in the Karanganyar district, where 67 people were killed in a landslide on a mountain slope. The victims were buried in mud up to 16 feet (5 m) deep. Local people used their hands and simple tools to rescue those trapped. In Java, at least 120 people died in the landslides.

News Flash

Thursday, December 27, 2007

Rescue teams in Indonesia are trying to reach dozens of people buried in landslides on the island of Java after floods and landslides left nearly 100 people dead or missing. Officials say landslides hit villages in densely populated central Java's Karanganyar and Wonogiri districts early on Wednesday.

• Mud and rocks dumped by the Java landslides buried this house nearly to its roof.

What Causes Landslides?

Weathered pieces of rock slide down a mountain when they break off. Usually, the pieces collect in one place until rains, melting snow, or an earthquake start them moving downward. Sometimes this happens at great speed.

Rock Falls

Pieces of rock, or scree, break away from the mountain and collect at the foot of the slope. After many years, they form a cone-shaped pile, called a scree slope. There are many small **rock falls** on the slopes of a mountain. They are most common in the spring when ice that has formed in cracks in the rocks begins to melt and releases the pieces of rock. They also occur after a heavy rain has soaked the soil or rock.

- These three kinds of landslides show how rocks and soil can move down a slope.

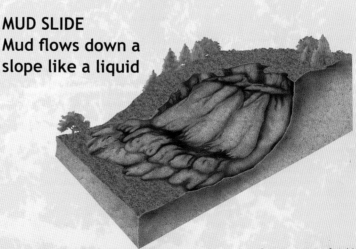

ROCK FALL
Falling scree slowly forms cone-shaped piles of rock

SLUMPING
Slipping blocks of rock form steps

MUD SLIDE
Mud flows down a slope like a liquid

Slumping

When a huge block of rock falls during a landslide, it is called slumping. Often set off by an earthquake, slumping is most common where a layer of rock, with many cracks in it, lies on top of a smoother, slippery rock such as clay or **shale**. The slumping blocks of rock may be up to 2 miles (3 km) long and 500 feet (150 m) thick. Such a slump often leaves large steps down the side of a mountain.

Mud Slides

Mud slides usually occur after a heavy rainstorm or when snow melts rapidly on a mountain. The soil or other loose material then becomes soaked with water. Some mud slides are caused by water mixing with the ash from a volcano. The wet material no longer sticks to the slope and slides downward, flowing like a liquid.

DATA FILE

- The digging of mines and quarries in mountains, or forest clearing from mountain slopes, can start a landslide. They can also make one worse than it would have been.

- Mud slides can travel at speeds greater than 47 miles (75 km) per hour.

- In 1920, a series of landslides in China killed 100,000 people and destroyed many villages.

- Rock falls are most likely to occur when ice melts higher up a mountain, allowing weakened rocks to fall.

Deadly Avalanche

In 1999, a huge avalanche struck the small town of Galtur in the Austrian Alps. Galtur was buried under 16 feet (5 m) of snow, and 31 people were killed.

Hidden Block of Snow

January 1999 was a warm month in the Austrian Alps. But this was followed by record snowfalls and strong winds in February. The people of Galtur did not know that a huge and unstable block of snow was forming high up on one of the mountain slopes above the town.

- The inhabitants of Galtur had little warning of the devastating snow avalanche.

Name: Galtur avalanche
Location: Galtur, Austria
Type of disaster: Avalanche
Deadly event: February 1999
Fatalities: 31

Lucky Survivors

On February 23, rising temperatures and showers of rain loosened the block of snow. The 187,000 ton (170,000 t) block began to move down the slope. It picked up speed until it reached nearly 186 miles (300 km) per hour. It took less than a minute to hit the valley below and double in size. This **avalanche** buried Galtur under 16 feet (5 m) of snow. Houses and buildings were crushed, roads were blocked, and telephone lines were brought down. For the night, rescue helicopters could not reach the disaster zone because of raging snowstorms. Amazingly, 26 survivors were dug out alive.

News Flash

February 24, 1999

At least eight people have been killed after an avalanche hit the small town of Galtur in the Alps in western Austria. It is thought that more than 25 others are still trapped under snow, but even before night fell, the rescue operation was being hampered by bad weather and fading light.

• A specially trained dog searches for avalanche victims. It takes a dog half an hour to search an area that would take 20 people 4 hours to cover.

What Causes Avalanches?

An avalanche is a mass of ice or snow that suddenly crashes down the side of a mountain into the valley below. Many avalanches sweep along chunks of rock, which makes the avalanche even more dangerous.

Powder Avalanche

There are two main kinds of snow and ice avalanches. The first kind is a **powder avalanche**. This type of avalanche can occur in very cold, dry weather when light powdery grains of snow do not stick together. If this snow starts to move down the mountain, it forms a powdery mass of snow that swirls along like an enormous white cloud.

• A huge powder avalanche occurs on Alaska's Mount McKinley.

28

Slab Avalanche

A **slab avalanche** starts as a solid chunk of frozen snow that can be as large as a football field and 30 feet (9 m) thick. It often forms when sunny days are followed by frosty nights. This change in temperature causes melted snow to freeze again. When this massive slab suddenly starts sliding down a mountain at great speed, it often carries huge rocks and trees with it.

Avalanche Wind

When an avalanche rushes down a mountain, a wind is produced ahead of the mass of snow or ice. This is known as an avalanche wind. This wind is often strong enough to demolish entire buildings.

DATA FILE

- An avalanche can range in speed from 62 miles (100 km) per hour to 244 miles (392 km) per hour.

- There are up to 1 million avalanches worldwide each year that kill about 100 people.

- To help prevent avalanches, wood, aluminum, or steel fences are built on steep slopes where snow collects in deep drifts.

- Guns or small explosives are sometimes used to set off safe, harmless avalanches.

- Forests of large, mature trees growing on mountain slopes form one of the best protections against avalanches.

- A destructive avalanche wind will rush down a mountain ahead of an avalanche such as this.

Glossary

avalanche a mass of snow, rock, and ice suddenly sliding down a mountain

block mountains mountains that have been formed where land has been pushed up close to a fault or between two faults

crust the outer layer of Earth, made up of huge slabs of rock

dome mountain a shallow, rounded mountain formed from liquid rock that is pushed up under Earth's crust

erosion the natural wearing away of land by wind, moving water, or ice

fold mountains mountains that have been pushed into folds or ridges

by movements of Earth's plates

glacier a large river of slow-moving ice, or ice and rock, that forms in a mountain and moves very slowly down a valley

landslide soil or rocks sliding down the side of a hill or mountain

magma the hot, molten rock formed in Earth's mantle below the crust

meltwater water that is released from melting snow and ice

oxygen colorless, odorless, and tasteless gas that is essential for the survival of most living creatures

peak the pointed top of a mountain

plates sections of Earth's crust that fit together like pieces of a giant jigsaw puzzle

powder avalanche an avalanche made up of powdery grains of snow that do not stick together

range a row, or line, of mountains

rock the solid part of Earth's crust beneath the soil

rock fall fragments of rock, or scree, that break away from the face of a steep cliff, hill, or mountain and collect at the foot of the slope

scree pieces of weathered rock that collect at the bottom of a steep mountain slope

shale a soft rock made of solid mud

slab avalanche an avalanche that consists of a huge slab of frozen snow

summit the top of a mountain or hill

sulfur dioxide a smelly gas that is released from volcanoes, especially during eruptions

valley a low-lying strip of land between steep hills or mountains

volcano a hole or tear in Earth's crust from which molten rock flows

weathering the breaking up of rocks by heat, cold, ice, and rainwater

Index

Web Sites

http://www.nationalgeographic.com/ngkids/0301/
This site provides stories, tips, facts, and a video of an avalanche.

www.mountain.org/education/explore.htm
Discover how mountains are made, their secrets, and their folklore.

http://www.factmonster.com/spot/everest2.html
Learn about people who have climbed Mount Everest.

http://www.kidsgeo.com/geology-for-kids/0071-landslides.php
This site covers a variety of topics including landslides and erosion.